INTERNATIONAL CENTRE FOR MECHANICAL SCIENCES

COURSES AND LECTURES - No. 7

VLATKO BRČIĆ
UNIVERSITY OF BELGRADE

APPLICATION OF HOLOGRAPHY AND HOLOGRAM INTERFEROMETRY TO PHOTOELASTICITY

LECTURES HELD AT THE DEPARTMENT
FOR MECHANICS OF DEFORMABLE BODIES

2nd ed.

UDINE 1974

SPRINGER-VERLAG WIEN GMBH

ISBN 978-3-211-81163-4   ISBN 978-3-7091-2646-2 (eBook)
DOI 10.1007/978-3-7091-2646-2

CONTENTS

LIST OF SYMBOLS USED

$\bar{\pi}$	position vector in the space of the holography appa ratus
$\bar{\rho}$	region of $\bar{\pi}$ - space occupied by the refractive mate rial of concern
$\bar{\pi}_o$	location of (an image or object point for) the detector which examines the recontructed image
$\bar{\bar{n}}\,(\bar{\rho})$	The refractive index distribution
$F(\bar{\pi},\bar{\pi}_o)$	fringe system in the reconstructed space
$F_N(\bar{\pi},\bar{\pi}_o)$	the N-th component of $F(\bar{\pi},\bar{\pi}_o)$
$I\,(\bar{\pi})$	a signal (electromagnetic, mechanical, chemical,...) which influences $\bar{n}\,(\bar{\rho})$
$\bar{A}(\bar{\pi},t)$	a vector wavefield complex amplitude
$A_n\,;\,\mathcal{U}_n\,;\,\mathcal{U}$	scalar wavefield complex amplitudes
$A'_n(\bar{\pi},t)$	a replica of $A_n(\bar{\pi},t)$
$\bar{\bar{B}}\,(\bar{\rho})$	a matrix representation of the effect of $\bar{\bar{n}}\,(\rho)$
$p\,(\bar{\rho})\,;\,q\,(\bar{\rho})$	transverse principal stress vector components
$\bar{k}_n$	propagation vector for $A_n(\bar{\pi},t)$
$\hat{\imath}_n\,;\,\hat{\jmath}_n$	unit vectors forming orthogonal triad basis with $\hat{k}_n$

δ_N retardation between the polarization components of the Fourier component, $\bar{A}_N (\bar{\varkappa}, t)$, in absence of the model

$\delta_2 (\bar{\rho})$ retardation between polarization components of $\bar{A}_o (\bar{\varkappa}, t)$

$\delta_m (\bar{\rho})$ differential retardation due to passage of $\bar{A}_I(\bar{\varkappa}, t)$ trought the model

$\delta_{2_n} (\bar{\rho})$ the $\delta_2 (\bar{\rho})$ for the n-th exposure

$\phi'_N (\bar{\varkappa})$ the phase of $\bar{A}_I (\bar{\varkappa}, t)$ at the hologram, in absence of the model

$\phi_N (\bar{\varkappa})$ the phase of $\bar{A}_o (\bar{\varkappa}, t)$ at the hologram, with the model present

$\phi_{N_n} (\bar{\varkappa})$ the $\phi_N (\bar{\varkappa})$ for the n-th exposure

$\Delta \phi$ the net single refractive component of the effect of the model on $\bar{A}_I (\bar{\varkappa}, t)$

$\phi_{NM} (\bar{\varkappa})$ the relative phase between A_N and A_M

$\phi_{NM_n} (\bar{\varkappa})$ the $\phi_{NM} (\bar{\varkappa})$ for the n-th exposure

$\bar{\omega}_{MN}$ the spatial frequency difference between $\bar{A}_M$ and $\bar{A}_N$

$a_N ; b_N$ relative amplitudes of the polarization components of $\bar{A}_N$

$H\,(\bar{\varkappa})$ the hologram recording

$H_{N}\,(\bar{\varkappa})$ the N-th component of $H\,(\bar{\varkappa})$

$F_{1}\,*\,F_{2}$ the convolution of F_{1} and F_{2}

$F_{1}\,*\,F_{2}^{*}$ the convolution of F_{1} and the conjugate of F_{2}

P_{m} either of the two orthogonal polarization states for the m-th exposure

$\Pi_{m}\,;\,\Sigma_{m}$ one of the polarization states, and the m-th exposure

α_{n} the front face of the n-th "in effect" reconstructed model

β_{n} the rear face of the n-th model, in the ensemble of reconstructed images.

PREFACE

This text represents a short introductory view in ho
lography and hologram interferometry and their possible applica
tions to experimental stress analysis, particularly to photoelasticity, without pretending of being a complete presentation
of this matter.

The experimental results are analyzed for three proto
type holograms recorded of a plane transparent specimen subject
ed to mechanical load. The analysis provides a comparison of the
methods and results obtained in hologram interferometry with the
corresponding experimental methods and results of polariscopy
and classical interferometry.

Few papers have been published on this matter; the
theoretical and experimental backgrounds and the explanation of
the phenomena which appear at the application of hologram inter
ferometry to photoelasticity have been still in developmental
stage. The main idea of the presented text originate as the re
sults of a team work being carried out by R.L. Powell, J.D.
Hovanesian, and the author of this text in Ann Arbor, Michigan,
during 1966 and 1967.

It is my very pleasant obligation to express the grat
itude to the Secretary General Prof. Luigi Sobrero and to the

Rector Prof. Waclaw Olszak for inviting me to present this course

at the International Centre for Mechanical Sciences in Udine.

September, 1969

V. Brčić

INTRODUCTION

During the past decade there has been renowed interest in Holography, in Gabor's well-known wavefront reconstruction process[1] *, the method of storing information concerning the three-dimensional nature of an object. In Holography, one records the electro-magnetic field that has been reflected from some object scene. By interferometric methods to create standing wave patterns in space, one can record the entirety of this wave pattern, both amplitude and phase, on some medium such as suitable photographic film, which responds to intensity only. At some arbitrary later time one can recreate the original wave pattern by illuminating this photographic record, called a hologram, with a beam of coherent light. One then sees the replica of the original object scene, in full three-dimensional form. This replication is, to all appearances, indistinguishable from the original scene.

This process can in principle be carried out at any wavelength, from microwave and infrared to X-rays and gamma rays. The wave need not to be electromagnetic. One can record waves of one frequency and make the reconstruction at another, for example, record electron waves and reconstruct using vis-

---

* [.....] References

ible light.

The field of holography is relatively new, the
G bor's paper appeared in 1948, but because of the relatively low
spatial coherence of available light sources at the time, its
applications were very limited and few investigators were activ
in this field. The recent discovery of lasers, particularly the
current availability of high-intensity lasers which posses desir
able spatial coherence, has permitted and encouraged extensive
work in various phases of holography.

A series of papers has appeared recently giv-
ing details and informations about the possibilities of this new
physical branch (see papers by Leith, Kozma, Upatknies, and oth.
[2],[3], the book by Stroke[4], by Pauthier-Canier[5], and many
others).

The application of holography in experimental
stress analysis has been stimulated by hologram interferometry
particularly. The development of hologram interferometry has been
dated last few years in papers by Stetson and Powell[6],[7],[8],
Hildebrand and Haines [9],[10], and others, offering the possibi
lity to applications in photoelasticity, strain analysis, vibration phenomena, etc. The first data on the applications of holography in photoelasticity were given in papers by Fourney[11],
Hovanesian, Brcic and Powell[12],[13],[14], Nicolas[15], and some

others.

The brief presentation of holography and hologram
interferometry and their applications in photoelasticity is the
subject of the following text. The accentuation will be given to
the analysis of three prototype holograms recorded of a plane
transparent specimen, under an applied mechanical load. The anal
ysis provides a comparison of the methods and results obtained
in hologram interferometry with the corresponding experimental
methods and results of polariscope and classical, single-stage-
-of-diffraction interferometer instruments. The analysis exhibits
the source of certain inherent advantages of the hologram interferometer over the polariscope and classical interferometers.
Finally, some speculative indications are presented of the power of hologram interferometry for handling more general engineer
ing problems than the prototype discussed.

2. BASIC CONCEPTS IN HOLOGRAPHY AND HOLOGRAM INTERFEROMETRY

To understand how a different kind of lightning can be used in holography, in "lensless photography", let us look at the nature of light in a little more details. Ordinary light is made up of many different wavelengths, none of which maintain a fixed phase relationship with each other, or with themselves, over a period of time. Saying in another way, it has poor temporal coherence; the waves at one point in the beam do not remain

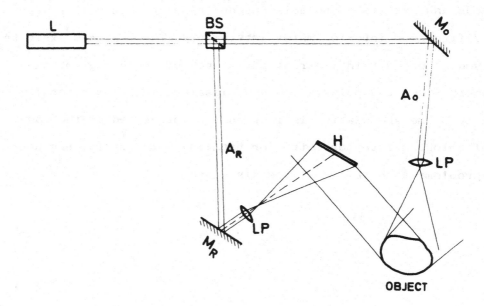

Fig.1 - Schematic Presentation of Hologram Formation

L	Laser	A_R	Reference Beam
BS	Beam Splitter	A_O	Object Beam
M_O, M_R	Mirrors	H	Holographic Plate

in step with the waves at some point in the beam further up
stream. Such incoherent light does not scatter from objects in
an orderly manner, and, most important for the application to
holography, is not capable of interfering with itself.

However, lasers produce the light beams which are
coherent over 10^{10} wavelengths and more. Such sources are almost
universally used for experiments in holography. The fact that
the light from a laser remains in step with itself along the
length of the beam is part of the reason, but the beam's intensity, like produced by lasers, is almost equally important.

The process of holography consists of two steps.
The first step, called the hologram formation, is shown schemat
ically in Fig.1. The laser beam is divided by the beam splitter
and two beams are reflected by mirrors M_0 and M_R and directed
so that they impinge upon the hologram plate. It is necessary
that the difference of the total path length of separate beams
be within the spatial coherence length of the source. These two
beams are called "object beam" and "reference beam", respective
ly. The object of which the hologram is to be formed is illuminated by this coherent light and after being scattered by the
object combines with a second beam of coherent light ("reference
beam"), to form an interference pattern which is recorded on a
photographic plate. This plate, after normal photographic proc

essing, contents the information required for the second step of
the process, the wavefront reconstruction.

A hologram is therefore an interference pattern
between a reference wave and the waves scattered by the object
being recorded. The result is a kind of diffraction grating modulated by the particular information contained in the object
being recorded. In essence, the plane wave from the laser acts
as a sort of carrier wave which is modulated by the signal from
the object. This signal is separated from the carrier by compar
ing it with unmodulated wave, and this "demodulated" signal is
then recorded.

The unexposed hologram plate is a spectrographic-
-type transparency emulsion (Kodak 649F) which possesses high
resolution. After exposure and convential photographic processing with developer D-19, the hologram appears as a blurred unin
telligible exposure which seems to bear little or no relationship to the original information. Fig.2. represents the photography of a hologram.

The information recorded in a hologram is actual-
ly much greater than the information in a photograph. The hologram records not only the lightness at a certain point in a scene,
but also the fore-and-aft location of the point and its color.
It records, in fact, all the informations about the object that

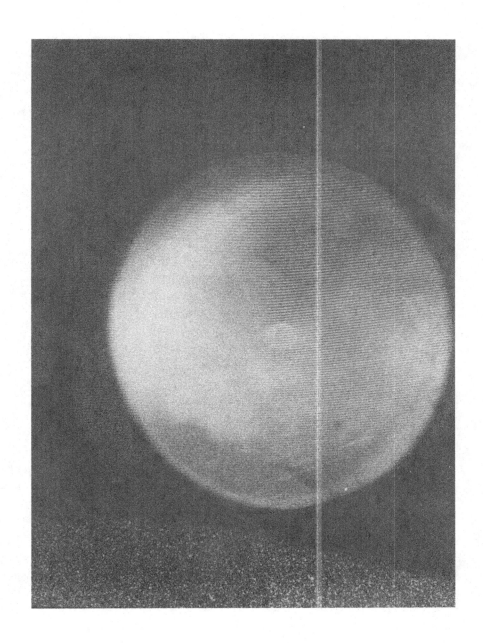

Fig. 2 - Photograph of a Hologram

can be conveyed by scattered and reflected light. Hence, the name "hologram" from the Greek root for "whole" and "writting".

From the point of view of optics, the hologram is recognized as an interferogram; it is a record of the pattern of mutual interference between the two radiation fields. From the point of view of electrical engineering and communication theory, the recording process achieves the modulation of the complex spatial frequency content of the object beam, onto the spatial frequency carrier, the reference beam, and stores this modulation pattern as the recording.

The hologram of Fig. 2, when illuminated by cohe-

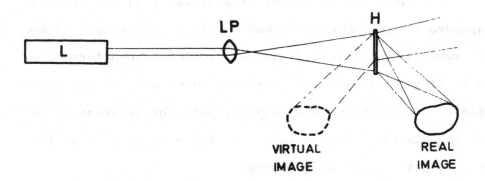

Fig.3 - Schematic Presentation of Hologram Reconstruction Process

rent monochromatic light, can render a reconstruction of the
original object. This is the second step in holography called
holography reconstruction process. Upon illumination by a coherent light source (in practice this is the reference beam), light
is diffracted by the hologram to form the images of the considered object. The first-order diffraction pattern forms both
a real and virtual image of the object (see Fig. 3). Without
entering now into the physical explanation of the reconstruction process, let it suffice to say that, upon illumination, the
high-frequency interference patterns in the hologram produce a
systematic difraction pattern which regroups the light rays into
pattern which depicts the original object.

Thus, to an observer the properly illuminated holo
gram appears to be a window through which is seen a three-dimen
sional image of the object. A virtual image has been formed, by
reconstructing the object beam one the output side of the hologram plate. Relative to the hologram plate, the apparent position of the virtual image is precisely the same as that of the
object during the reconrding process.

Hologram interferometry is an extension of holography and it may be accomplished in different ways. One method is to place the processed hologram exactly in the same position it occupied during exposure, thence to subject it to the

same two beams, with the model in place. If the model is exactly in the same state and position as it was originally, one
would observe an identical superposition of the virtual image
of the object and the object itself, not being able to distin-
guish one image from the other. If the object is subjected to
some small change in position or if it is subjected to some
loads which cause optical changes in the material, a system of
interference fringes is produced which is essentially identical
to those observed in a Michelson or Mach-Zehnder interferometer.

Stetson and Powell[8] have shown that the same
effect can be produced on a single hologram by subjecting the
emulsion to two exposures. The first exposure is made with the
object in some original state and position. The second exposure
is of the same duration and is made with the object in some
other state or position, say,as a result of applying forces to it.
This single double-exposed hologram will render a reconstruction of the image along with families of interference fringes
which depict the differences of the two states. The above is
possible because of the experimentally proven principle by
Stetson and Powell, that the resulting image reconstruction is
unaffected due to whether the events occured simultaneously or
sequentially during the hologram exposure (formation, not re-
contruction) period.

The fringes which appear at such a process are due to interference between the two nearly identical reconstructed signal beams. It is possible to relate both qualitative and quantitative features of the fringes to displacements which occur in the interferometer during the recording process. The study and exploitation of the information content of the fringes is a part of what has come to be called hologram interferometry, wavefront reconstruction interferometry, holographic interferometry, or Moiré technique in holography.

The wavefront reconstruction process, together with multiple exposure techniques, can generate in the reconstructed scene inetrference fringes whose orientation, spacing and location can furnish data related to local displacement in the apparatus.

3. APPLICATION OF HOLOGRAPHY TO PHOTOELASTICITY

3.1. Introductory Considerations

The holographic image of a birefringent object is not transparent. The reconstructed image scene includes a three-dimensional distribution of visible interference effects, in addition to, and superposed upon the replicated array of illuminated object elements. This circumstance contains two reciprocally related implications which provide the theme of this discussion:

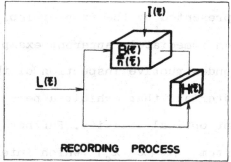 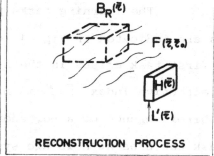

Fig.4 - Two Tasks of Hologram Interferometry with respect to Birefringent Objects:
(1) To Apply Hologram Interferometry to an Engineering Problem - given $F(\bar{\varkappa}, \bar{\varkappa}_0)$; to specify $B(\bar{\varkappa})$
(2) To Develop the Rules of Operation of the Interferometer - given $B(\bar{\varkappa})$; to specify $F(\bar{\varkappa}, \bar{\varkappa}_0)$.

(1) when the pattern of birefringence in the object space can be inferred from a knowledge of the interference display in the image space, a possible engineering application of holography is indicated.

(2) on the other hand, when qualitative and/or quantitative features of the interference pattern can be inferred, given a known distribution of birefringence, then the operational rules which govern the apparatus, as an interferometer, are exhibited.

The learning tasks presented by the pair of impli cations are indicated in Fig. 4. An immediately apparent example of the first task arises in the nondestructive inspection of the local refractive index $\bar{\bar{n}}(\bar{p})$ in materials that exhibit a permanent birefringence or a permanent optical activity. Further, this task can present a fringe system $F(\bar{x}, \bar{x}_o)$ which inter rogates temporary incremental disturbances of the field due to a controlled external influence $I(\bar{x})$: this may be, e.g., a thermal, mechanical, chemical, or electromagnetic disturbance, which varies essentially discretly, periodically, impulsively or stochastically. In such situations the process can employ an influence of known properties, with the aim being to inspect the response of the material, $\bar{\bar{n}}(\bar{p})$. Reciprocally, it can employ a material of known $\bar{\bar{n}}(\bar{\rho})$ around which to design an

instrument to detect the presence of, or, moreover, the informa-

tion content in the influence $I(\bar{\lambda})$.

Exploitation of the wealth of engineering uses ex-

emplified by task 1, however, requires a careful attention to the

work of the task 2. Inherently the wavefront reconstruction pro-

cess provides an abundance of physical information. This is

certainly the case, also, with <u>hologram interferometry</u>. In much

to same manner that the visual image replication by holography

is intrinsically more informative then object replication by

photography, so is hologram interferometry (HI) more profilic in

its quality and quantity of information read-out than it is clas

sical, single-step interferometry process (CI). For example, not

only can the HI process develop obviously meaningful fringe sys

tems related to structures which can not be examined by CI in-

struments $[6]$, $[10]$; it can, as well, easily generate fringe systems

the interpretation of which overtaxes the teachings of fringe

formation and fringe location that have been developed with care

during the course of the growth of CI as a branch of Optics and

as a tool for use in engineering problems.

If to work at the task 1 requires attention to task

2, and since the latter presents the optics questions whose di-

rect analogues do not already exist in the prior art of the parental disciplines, then it is clear that success in general with the applying of HI to engineering problems requires a cautious examination of the similarities and differences between the optical procedure and all other instruments, procedures, and concepts which contribute a parental component. In order to proceed most efficiently with the task 2, as a support to task 1, there is a minimum list of categories of classical disciplines, concepts, and methods which must be drawn upon: interferometry, polariscopy, schlieren optics, communication theory, diffraction theory. Considered within this formidable framework, the problem which we discuss here is a modest entry to the first task; further, it employs the resource disciplines in only a rudimentary manner. It does, however, permit some incremental insight into the power of HI as an optical instrument; it does exemplify the value of the HI process in terms of its usefulness as an engineering tool for problems more general than the one discussed here.

3.2. Experimental Arrangements

We undertook the problem of the holographic study
of plane plastic sheet specimens under the influence of mechanic
al loads applied perpendicularly to the thickness surfaces of
such models $[12]$, $[13]$. The distribution of temporary birefringence
produced by the load, $B(\bar{\rho})$, is of a form tipically studied
by a variety of polariscopes (P) and CI techniques in photoelastic
stress analysis work with two-dimensional models. This choice of

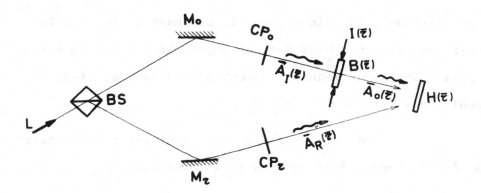

*Fig.5 - Optical Bridge Circuit for a Representative
Hologram Recording Apparatus*

L *—CW Laser Beam Input to the Interferometer*

BS *—Beam Splitter*

M_o, M_R *—Mirrors in Object Beam and Reference Beam*

CP_o, CP_r *— Circular Polarizers*

$B(\bar{\rho})$ *— The Birefringent Object*

$I(\bar{F})$ *—Externally Applied Mechanical Load*

$H(\bar{\chi})$ *— The Hologram Plate*

"object" for the holography permits the interpretation of the
consequent $F(\bar{x}, \bar{x})$ to benefit from the comparison with
the similar fringe data produced in photoelasticity instruments.

A representative arrangement of the apparatus is
depicted in Fig. 5. The laser employed can be a He-Ne continuous, f.e., 50 mw source, with wavelength 632.8 mm.
Brewster angle exit windows are a feature of the laser; the
radiation input to the interferometer is therefore linearly polarized. In order to be able to exploit the teachings of polariscope instrument techniques, provisions have to be made to
convert the reference beam and the object irradiance to circular polarization, by means of essentially identical elements
(circular polarizers) CP$_r$ and CP$_o$.

From Fig. 5 it is clear that the basic interroga
tion of the object $B(\bar{p})$ occurs in an operation

$$\bar{A}_o(\bar{x}) = \bar{\bar{B}}(\bar{p}) \cdot \bar{A}_I(\bar{x}) \qquad (1)$$

by which the complex amplitude $\bar{A}_I(\bar{x})$ of the object illu
minating field is transformed into the complex amplitude $A_o(\bar{x})$
of the "object beam" for the hologram recording process. Let us
pose the question, can the fact that the reconstructed object
space, unlike the illuminated object space, is not transparent,

. i.e., the fact that the fringes exist in the space of the former,
be accounted for by means of an analytical examination of the
hologram, i.e., the recorded interference pattern, for which
equation (1) is the object beam. The answer to this question
will be given in the next Section. First, however, let us recall
some of basic results of experimental investigations utilizing
an apparatus such as Fig. 5.

3.2.1. Three Prototype Experimental Results

Consider a specimen so prepared as to have neglig
ible residual stress, i.e., induced birefringence under a condi
tion of no applied load. The model is located in a supporting
structure which can apply a system of external forces, as describ
ed above. Now, let three holograms be recorded. In the first,
the "object" is the model under no load. The second hologram is,
again, a "single exposure" recording for which the"object"is the
model, under the terminal load used at the experiment. In the
third hologram, a "double exposure" hologram is recorded: an
exposure is made while the model is unloaded, the light input to
the interferometer is obstructed and a system of forces is applied to the model; the light input is then readmitted, continu-
ing the exposure of this third hologram plate. When this set of
holograms is processed, and interrogated by a replica $\bar{A}'_R (\bar{x})$

of the reference beam, the images observed exhibit the following features. The first image is transparent; each of the second
and the third image gives a system of fringes, in addition to
the replica of the model. The analysis of these fringes shows
that the fringes in the image from the second hologram are isochromatics, contours of the constant difference of the transverse
principal stress vector components, p $(\bar{p})$ and $q(\bar{p})$, and the
fringes in the image of the third are a distinguishable composi
tion of the isochromatics and isopachics (contours of the cons
tant sum of the transverse principal stress vector components),
for the observed location $\bar{p}$ on the model. In this latter case,
the image from the third hologram, the observable may be describ
ed, crudely as a display on one of the fringe families, but with
its visibility modulated by a distribution which is itself a re
cord of the disposition of the other family. Fourney[11] , Hovanesian, Brcic, and Powell[12] , [13] have pursued this approach
independently, outlining, via the stress-optics laws, the corre
lation between the fringe structure and local values of the prin
cipal stresses. In the light of the foregoing experimental ob-
servations, the primary task of our discussion can be stated, it
is basically two-fold:

        (1) to construct an **analytical** framework in terms
        of which to express, as a question in hologram

interferometry, how does the optical information of engineering interest come to be stored into the recorded interference pattern;

(2) within the expressions of (1), to indicate, as an example of the applications usefulness of the HI process, what implications are contained in the analytical expressions, relevant to visibility properties of wave fields reconstructed from such a hologram. In particular, we seek information in a language meaningful to the applications engineer; moreover, we seek, in the implications, suggestions as to how the HI apparatus might be applied to problems more general than our example.

3.2.2. Notes on Experimental Arrangements

In order to form hologram it is necessary that the model, the photographic plate, and the two beams remain stationary relative to each other during the required exposure time[11] . Since the diffraction pattern being formed on photographic plate has a spacing of the order of the wavelength of the light being used, the relative motions of these items that is tolerable is limited to values less than this spacing. In order

to achieve this, a heavy marble or metal platform, cca 1.50x2.
00x0.20 m, and properly isolated from vibrations has to be used.
In the paper by Fourney[11] the following recomandations to the
practical achievement have been given:

> (1) The path lengths of the refernce and object
> beams should be made approximately equal so that
> the spatial coherence of two beams will be maintain
> ed.

> (2) The intensities of the two beams to be made
> approximately equal at the photographic plate so
> that the density gradients of the recorded inter-
> ference pattern will be maximized.

> (3) The angle between the two beams at the holog-
> raphic plate be made as small as possible.

The first requirement depends on the type of laser
being used (usually it is a continuous He-Ne laser, output
power 20-50 mw), while the second and the third requirement de-
pend on the type of film being used. Violations of these require
ments will result only in a degradation in the quality of the
obtained hologram.

A tipical holographic polariscope equipment, as

Fig. 6   Typical Equipment Arrangement
a - Laser;   b - Model and Loading Device;
c - Holographic Plate;   d - Beam Splitter;
e - Mirror;   f - Lenses;   g - Pinhole and
Positioning Device for Spatial Filtering;
h - Light Diffuser and Polarizer;   i - Polar-
izing Element for Reference Beam;   j - Object
Beam;   k - Reference Beam.

given by Fourney[11] , is presented in Fig. 6.

          The standard classical polariscope consists of
light source, polarizer and analyzer which have their specific
functions: the first, to produce light of specific state of polarization, and the second to act upon this light to produce the
desired intensity variations.

          In the holographic polariscope the same functions

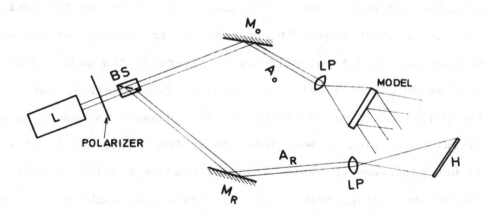

*Fig.7 - Schematic Presentation of Holographic
Polariscope.*

are performed by different way. Fig. 7 represents a schematic

arrangement of the holographic polariscope. Since the laser pro

duces plane polarized light, the polarizer is not necessary in

some cases. Since the hologram obtained with two coherent light

beams is a function of their relative polarization, the analyzer

is not necessary. It is also necessary to insert the pinhole and

positioning device for spatial filtering of both the reference

and the object beam.

Of course, some alterations are always possible,

with the aim to improve the quality of hologram and the quality

of reconstructed image. The contrast of the fringe pattern will

be enhanced if a light diffuser is placed in the object beam

just ahead of the model. Since this diffuser destroys partially

the polarization of the light, it is necessary to insert an addi

tional polarizing element into the system. Also, the light which

is not polarized properly, will not produce a hologram, but will

expose the photographic film. This extraneous light acts as a

background noise on the holographic diffraction pattern which

deteriorates the quality of the hologram. This is particularly

true when isoclinics are recorded. All light polarized at angles

other than the isoclinics angle is extraneous, hence, it is de-

sirable to eliminate this light by placing an analyzing element

in the object beam. However, this arrangement reduces the light

intensity and is used only when required. Fig. 8 represents the arrangement for obtaining isoclinics patter.

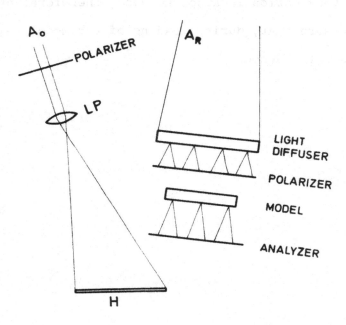

Fig.8 - Scheme of the Arrangement for Recording Isoclinic Fringe Patterns.

       The isopachics pattern, obtained by double exposu re (the third beforementioned prototype hologram) is the result of the change in optical thickness of the model due to applied external load. It is important to note here that only the change in optical thickness that occurs between the two exposures is recorded. Hence, no restrictions are placed on the model, such as flatness or paralel faces (what is, however, very important in classical interferometry).

The major limitation of the holographic method is the rigid-body motion is also recorded. Therefore, care must be taken to assure that, during loading of the model, extraneous motion is not induced.

3.3. The Analysis [14]

Let the complex amplitude of the reference beam field in the hologram plate be represented by

$$\bar{A}_R (\bar{x}, t) = \left\{ \hat{\imath}_1 \cdot 1 \cdot e^{i \bar{k}_1 \bar{x}} + \hat{\jmath}_1 \, b_1 \, e^{i(\bar{k}_1 \bar{x} + \delta_1)} \right\} e^{-i\omega t} \qquad (2)$$

The $\bar{A}_R(\bar{x}, t)$ is taken by (2) to be a generally polarized plane wave of infinite extent transverse to its propagation direction $\hat{k}$. The relative amplitude of orthogonal polarization components in the conventionally chosen transverse directions $\hat{\imath}_1$ and $\hat{\jmath}_1$, is b_1. The retardation between the two components is δ_1. The optical frequency time dependence, contained in the factor $e^{-i\omega t}$ will often be assumed tacitly hereafter, where appropri_ ate, and not carried explicitly except when its presence as a time dependence is to be emphasized. A more realistic expression for (2) might include the finiteness of transverse spatial extent, and diffraction limited decollimation. However, for the gross features we seek to exhibit, such an integral formulation is unnecessary mathematical equipment.

With similar interpretation, the complex amplitude of the optical field incident on the model is represented by

$$\bar{A}_I (\bar{x}, t) = \left\{ \hat{\imath}_2 \, a_2' \, e^{i \bar{k}_2 \bar{x}} + \hat{\jmath}_2 \, b_2' \, e^{i(\bar{k}_2 \bar{x} + \delta_2')} \right\} e^{i \phi_2(\bar{x})} \qquad (3)$$

Here, ϕ'_2 accounts for the phase of the field (3), relative to that of the reference beam, consequent to the choice of optical path trajectories between the location of beam division and the position of the model. In the design of an apparatus the worker has the choice of establishing the values for elements of the sets, $(\bar{k}_1, \bar{k}_2)$, $(\hat{\imath}_1, \hat{\imath}_2)$, b_1 , a'_2, b'_2 , and $\phi'_2(\bar{\kappa})$, in somewhat the manner that the electrical engineer can select the values of parameters for the devices and relationship which composed the lumped and/or distributed elements of the circuits which make up his instrumental systems for detecting, inspecting, and processing of electronic signals.

In anticipation of the task of comparing our expressions with P and CI expressions, let us describe the interaction of the specimen with (3) as a process which merely transforms the latter into a field whose complex amplitude, in the space beyond the model, is

$$\bar{A}_0(\bar{\kappa}, t) = \left\{ \hat{\imath}_2\, a_2\, e^{i\,\bar{k}_2\bar{\kappa}} + \hat{\jmath}_2\, b_2\, e^{i\left[\bar{k}_2\bar{\kappa} + \delta_2(\bar{p})\right]} \right\} e^{i\,\phi_2(\bar{\kappa})} \qquad (4)$$

Here, $\delta_2(\bar{p})$ represents the retardation in the "illumunated object" field, (4), which is presented to the hologram plate in the recording process. The modification of δ'_2 , namely to $\delta_2(\bar{p})$, is a partial characterization of the mechanical information of our example. The $\phi_2(\bar{\kappa})$, a_2 , b_2 , similarly are modifica-

tions which can contain information of applications interest.
The ϕ_2 is a phase factor which contains the influence, at a
location $\bar{p}$ on the model, of its net "singly" refractive nature. In the spirit of the P and CI instruments, these model
parameters are taken as local on the model surface. That is to
say, the inspection analysis assumes a two-dimensional model.
It is assumed that (4) contains the result of the performance
of an optical inspection integrated along the lines of sight
provided by (3). Hovanesian, Brcic, and Powell[12] and Fourney[11]
have indicated the analysis connecting the optical data of interest

$$\delta_m(\bar{p}) = \delta_2(\bar{p}) - \delta_2'$$

$$\Delta\phi_m(\bar{p}) = \left\{\phi_2(\bar{x}) - \phi_2'(\bar{x})\right\}_{\bar{x}=\bar{p}}$$

to the questions of concern, namely the distribution of principal stress vectors in the plane problem, via the usual stress
opticals laws. In this Section, however, we focus attention on
the details of the optical, rather than mechanical, aspect.
That is, concern is more with the optics of how the inspection
is made than with an applications engineering description of
what is inspected.

The interference pattern recorded in the wave-

front reconstruction process is the time average of the squared
modulus of the superposition of the complex amplitudes (2) and
(4). This is accomplished when the photosensitive phate is caus
ed to be exposed to both beams simultaneously. The work involv-
ed in creating the latent image of this modulus squared pattern
may be described by a time integration of the divergence of the
field of the complex Poyinting vector associated with the super
posed field $A_T(\bar{x}, t)$, at a given location in the emulsion, for
the time duration of the exposure. Thus, when the exposed plate
has been processed appropriatelly, the emulsion consists of a
three-dimensional mapping of the mutual interference pattern
characteristic of $\bar{A}_T(\bar{x}, t)$. This recorded pattern is the ho-
logrām. Ordinarily it has been sufficient to express this average as the square of the modulus of the sum of two scalar fields.
In our case, however, it is appropriate to describe the informa
tion from in the emulsion as the square of the modulus of the
sum of two complex vector wave fields, namely the product

$$H(\bar{x}) = \left\{ \left[\bar{A}_R(\bar{x}) + \bar{A}_0(\bar{x}) \right] e^{-i\omega t} \right\} \left\{ \left[\bar{A}_R(\bar{x}) + \bar{A}_0(\bar{x}) \right] e^{-i\omega t} \right\}^*$$

$$\text{(5)}$$

$$= \bar{A}_T(\bar{x}, t) \cdot \bar{A}_T^*(\bar{x}, t)$$

This can be written in the form

This can be written in the form

$$H(\bar{\kappa}) = 1 + b_1^2 + a_2^2 + b_2^2 +$$

$$+ 2 \left\{ (\hat{\iota}_1 \hat{\iota}_2) \, a_2 \quad \cos \left[\bar{\omega}_{12} \, \bar{\kappa} - \phi_{12}(\bar{\kappa}) \right] + \right.$$

$$+ (\hat{\iota}_1 \hat{\jmath}_2) \, b_2 \quad \cos \left[\bar{\omega}_{12} \, \bar{\kappa} - \phi_{12}(\bar{\kappa}) - \delta_2(\bar{p}) \right] + \qquad (6)$$

$$+ (\hat{\jmath}_1 \hat{\iota}_2) \, b_1 \, a_2 \cos \left[\bar{\omega}_{12} \, \bar{\kappa} - \phi_{12}(\bar{\kappa}) + \delta_1 \right] +$$

$$\left. + (\hat{\jmath}_1 \hat{\jmath}_2) \, b_1 \, b_2 \cos \left[\bar{\omega}_{12} \, \bar{\kappa} - \phi_{12}(\bar{\kappa}) + \delta_1 - \delta_2(\bar{p}) \right] \right\}$$

where

$$\phi_{12} = \phi_1 - \phi_2 , \; and \quad \bar{\omega} = \bar{k}_1 - \bar{k}_2 \qquad (7)$$

a vector spatial frequency difference component basic to each term of (6).

Let us choose $\hat{\iota}_1$ and $\hat{\iota}_2$ as mutually parallel, as well as perpendicular to $\hat{k}_1$ and $\hat{k}_2$. Consequences of this selection are

$$\hat{\iota}_1 \hat{\jmath}_2 = \hat{\iota}_2 \hat{\jmath}_1 = 0 \; ; \qquad \hat{\jmath}_1 \hat{\jmath}_2 = \hat{k}_1 \hat{k}_2$$

Thus the pattern (6) is seen to be

$$H(\bar{\kappa}) = 1 + b_1^2 + a_2^2 + b_2^2 +$$

$$\qquad (8)$$

$$+ 2 \left\{ 1 \quad a_2 \; \cos \left[\bar{\omega}_{12}(\bar{\kappa}) - \phi_{12}(\bar{\kappa}) \right] + \right.$$

$$\left. + (\hat{k}_1 \hat{k}_2) \, b_1 \, b_2 \cos \left[\bar{\omega}_{12}(\bar{\kappa}) - \phi_{12}(\bar{\kappa}) + \left\{ \delta_1 - \delta_2(\bar{p}) \right\} \right] \right\}$$

When (8) is considered in the light of the basic scalar field
descriptions of the holography process, some useful interpreta
tions are apparent immediately. The recording is essentialy a
pair of fringes, each with the basic spatial frequency $\bar{\omega}_{12}$.

$$H(\bar{\kappa}) = \left[1 + a_2^2 + 1 \cdot a_2 \cos\left[\bar{\omega}_{12}\bar{\kappa} - \phi_{12}(\bar{\kappa})\right]\right] +$$

$$+ \left[b_1^2 + b_2^2 + b_1 b_2 \cos\left[\bar{\omega}_{12}\bar{\kappa} - \phi_{12}(\bar{\kappa}) + \delta_1 - \delta_2(\bar{p})\right]\right] \equiv$$

$$\equiv H_a(\bar{\kappa}) + H_b(\bar{\kappa}) \tag{8a}$$

As usual, the magnitude of

$$\bar{\omega}_{12} = \frac{2\pi}{\lambda}(\bar{k}_1 - \bar{k}_2)$$

expresses that the fringe separation, $\left(\left|\bar{\omega}_{12}\right|\right)^{-1}$, is proportion
al to the wavelength of the radiation employed, and is increased
when the angle between the propagation directions is reduced.
The direction of $\bar{\omega}_{12}$ indicates the three-dimensional spatial orien
tation of the basic fringe pattern, "the venetian blinds" sys-
tem, as the elemental pattern is often called. Thus, planes norm
al to the direction $\bar{\omega}_{12}$, are planes of constant optical density; the optical density in these planes varies sinusoidaly along
the $\bar{\omega}_{12}$ direction. Since this direction is normal to the direction of $(\hat{k}_1, \hat{k}_2)$, it is seen that the fringes (i.e. planes of
maximum optical density) are so oriented in the emulsion as to

bisect the angle $(\hat{k}_1, \hat{k}_2)$.

Each of the two fringe systems has a phase modula
tion. It is of interest to note that the engineering information
is in the hologram plate in the form of these phase modulations
of the elemental fringe system. In fact, that there $\acute{a}$re two dis
tinct systems is a direct result of there being two distinct
sources of modulation of the frequency $\bar{\omega}_{12}$.

With the foregoing interpretation, the operational
ly "single" exposure recording operation formulated in the Section 3.2.1. is in effect the simultaneous independent recordings
of the holograms of two nearly identical"objects". Thus the
second hologram of 3.2.1. is effectively a "double" exposure
hologram. Thus, we may expect to generate, in the reconstruction
process, two nearly identical, and nearly superposed image wave
fields. Since this pair will be reconstructed simultaneously and
coherently, it becomes reasonable to expect to observe visible
effects of mutual interference between the fields from $H_a(\bar{\varkappa})$ and
$H_b(\bar{\varkappa})$. The sketch of $F(\bar{\varkappa}_1, \bar{\varkappa}_0)$ of Fig.5 attemts to symbolize
these concepts. Parenthetically, let us note the relationship
between this argument and previous similar equivalence arguments,
as one of the most practical aspects of the usefulness of the
HI process. At the hologram interferometry experiment, like applied for studying of vibration phenomena[8], the HI process re
cords sequentially, and repetitively two essentially distinct

illuminated objects, the surface at each of its rest positions
under the periodic stress condition; the reconstruction process
reconstructs the two coherently, with the fringes as a conse-
quence. In the present situation, two essentially uncoupled, non-
-interacting objects are still recorded, albeit recorded simulta
neously.

The next step, therefore, is to deal with the
question, given that a replicated set of interfering fields is
to be expected, how do we infer, from the array of interference
effects recorded in the hologram, forms of the interference information in the replicated space, that are useful to the applications engineer, qualitative and/or quantitative mechanical
features of the birefringence and refraction distributions over
the specimen? This question constitutes the primary goal of the
next Section. Let us state this problem symbolically as

$$F \left(\bar{x}_1 , \bar{x}_0 \right) \Rightarrow \bar{\bar{n}} \left(\rho \right) \Rightarrow p_\tau \left(\bar{\rho} \right) ; q_\tau \left(\bar{\rho} \right) \qquad (9)$$

3.4. The Analysis Approach

In the examination of $F(\bar{\kappa}, \bar{\kappa}_o)$ we take an ad hoc construction which evolves by means of guidance from CI. The procedure is to examine analytically the expression (8) for the hologram plate; there, to ask what is the nature of the object(s) that has been employed to make the recording. This procedure derives from an interpretation of the equation given by Gabor in his classical paper on microscopy [16], namely

$$\mathcal{U} = \mathcal{U}_1 + \mathcal{U}_o - A_o\, e^{it_o} + A_1\, e^{it_1} = e^{it_1}\left[A_o + A_1\, e^{i(t_1-t_o)} \right] \quad (10a)$$

$$\mathcal{U}\,\mathcal{U}^* = A_o^2 + A_1^2 + 2\, A_o\, A_1\, \cos\,(t_o - t_1) \quad (10b)$$

Hence, U_o and U_1 are the complex amplitudes of two mutually coherent wave fields at the hologram plate. Consider that U_o is the Leith and Upatknies off-set reference wave field [17], and that U_1 is the field scattered from the object; in terms of information content about the object, we say this beam at the hologram plate is the object. The product $\mathcal{U}\,\mathcal{U}^*$ is essentially the hologram recording. Using (10a), (10b), and operational experience with the reconstruction process, we state the following reversible rule pair:

(1) If the object (wave field) is $A_1\, e^{it_1}$ as in (10a),

then its information (all we can expect to recover in reconstruction) is stored in the hologram in the amplitude and phase modulation as deplicted in (10b).

(2) If the hologram can be expressed in the form of (10b), and we assume knowledge of the reference beam geometry, etc, then the nature of the (unknown) object replicated field (10a) can be predicted by identifying those phase and amplitude modulations in (10b) which constitute the optical signature of the illumi nated object.

3.4.1. The Second Prototype Hologram

We now examine (8) in view of the second rule. In the absence of birefringence,

$$\delta_2 (\bar{p}) = \delta_2'$$

and in a recording situation for which $\delta_2' = \delta_2$, the H_a and H_b store identical phase modulation information about the object. To the extent that the information stored in the form of amplitude modulation is negligible, the two holograms become identical. In this circumstance, the rule instructs that the reconstructed images are spatially degenerate, indistinguishable.

Whatever features of the object that are characterized by the
phase modulation content in $\phi_{12}(\bar{\varkappa})$ will be reconstructed; i.e.,
if the model is scored, if the surfaces are not parallel, or
not plane, or contaminated by a non-birefringente oil or finger
print smear, then these features will merely be replicated in
the image. That is to say, in contrast with the CI situation,
in HI the foregoing types of extraneous "optical information"
do not result in fringes to complicate the study of birefringence: $F(\bar{\varkappa},\bar{\varkappa}_0)$ still degenerates to nothing.

Any influence which contributes a violation of
the degeneracy condition,

$$\delta_1 - \delta_2(\bar{p}) = 0 \tag{11}$$

however, will result in spatial distinction between the fringes
of H_a and H_b. Clearly, the birefringence of the specimen does
just this, in a manner which is informed according to (8). It
is of interest to note that the sufficient condition for system
atic quantitative detection of local birefringence is merely a
violation of (11); that is, the HI does not require, moreover,
the possibly more expensive condition, e.g.,

$$\delta_2' = \delta_1 = n \cdot \frac{\lambda}{A} \; ; \quad n = 1.000 \quad or \quad 2.000 \tag{12}$$

The disturbance of δ_2' by passage through the model results, then, in two distinct fringe systems in (8); thus two nearly identical, nearly superposed, mutually coherent images may be expected from the reconstruction process. The interference between this pair of replicated fields gives rise to the fringe data $F(\bar{x}, \bar{x}_o)$ in the image space.

3.4.2. The Third Prototype Hologram

In view of (8a) we can describe the hologram recording for the third of the three prototype holograms referred to above, as the incoherent superposition of the fringe patterns for the model under the two distinct conditions of loading.

$$H_3(\bar{x}) = \left\{ 1 + a_2^2 + 2\, a_2 \, \cos\left[\bar{\omega}_{12}\, \bar{x} - \phi_{12}(\bar{x})\right] \right\} +$$

$$+ \left\{ b_1^2 + b_2^2 + 2\,(\hat{k}_1 \hat{k}_2)\, b_1 b_2 \, \cos\left[\bar{\omega}_{12}\, \bar{x} - \phi_{12_1}(\bar{x}) + \left\{\delta_1 - \delta_{2_1}(\bar{\rho})\right\}\right] \right\} +$$

$$+ \left\{ 1 + a_2^2 - 2\, a_2 \, \cos\left[\bar{\omega}_{12}\, \bar{x} - \phi_{12_2}(\bar{x})\right] \right\} +$$

$$+ \left\{ b_1^2 + b_2^2 + 2\,(\hat{k}_1 \hat{k}_2)\, b_1 b_2 \, \cos\left[\bar{\omega}_{12}(\bar{x}) - \phi_{12_2}(\bar{x}) + \left\{\delta_1 - \delta_{2_2}(\bar{\rho})\right\}\right] \right\} \equiv$$

$$\equiv \left\{\left\{ H_2 \right\}\right\} + \left\{ H_c \right\} + \left\{ H_d \right\} \equiv$$

$$\equiv \left\{ H_a \right\} + \left\{ H_b \right\} + \left\{ H_c \right\} + \left\{ H_d \right\}$$

$$(13)$$

In the spirit of the ad hoc rule 2, the recon-

struction process is thus expected to present four nearly super
posed, nearly identical, mutually coherent "image" wave fields.
Interference effects may be expected between the six distinct
combinations, C(4,2), of the four elements, taken in pairs. The
phasis of this set of pairs of interference fringe systems are
readily seen to be

$$\cos\left[\bar{\omega}_{12}\,\bar{x}\,-\,\phi_{12_1}(\bar{x})\right]$$

$$\cos\left[\bar{\omega}_{12}\,\bar{x}\,-\,\phi_{12_1}(\bar{x})\,+\,\left\{\delta_1\,-\,\delta_{2_1}(\bar{p})\right\}\right]$$

$$\cos\left[\bar{\omega}_{12}\,\bar{x}\,-\,\phi_{12_1}(\bar{x})\right]$$

$$\cos\left[\bar{\omega}_{12}\,\bar{x}\,-\,\phi_{12_2}(\bar{x})\right]$$

$$\cos\left[\bar{\omega}_{12}\,\bar{x}\,-\,\phi_{12_1}(\bar{x})\right]$$

$$\cos\left[\bar{\omega}_{12}\,\bar{x}\,-\,\phi_{12_2}(\bar{x})\right]\,+\,\left\{\delta_1\,-\,\delta_{2_2}(\bar{p})\right\}]$$

$$\cos\left[\bar{\omega}_{12}\,\bar{x}\,-\,\phi_{12_1}(\bar{x})\right]\,+\,\left\{\delta_1\,-\,\delta_{2_1}(\bar{p})\right\}]$$

$$\cos\left[\bar{\omega}_{12}\,\bar{x}\,-\,\phi_{12_2}(\bar{x})\right]$$

$$\cos\left[\bar{\omega}_{12}\,\bar{x}\,-\,\phi_{12_1}(\bar{x})\,+\,\left\{\delta_1\,-\,\delta_{2_1}(\bar{p})\right\}\right]$$

$$\cos\left[\bar{\omega}_{12}\,\bar{x}\,-\,\phi_{12_2}(\bar{x})\,+\,\left\{\delta_1\,-\,\delta_{2_2}(\bar{p})\right\}\right]$$

$$\cos\left[\bar{\omega}_{12}\,\bar{x}\,-\,\phi_{12_2}(\bar{x})\right]$$

$$\cos\left[\bar{\omega}_{12}\,\bar{x}\,-\,\phi_{12_2}(\bar{x})\right]\,+\,\left\{\delta_1\,-\,\delta_{2_2}(\bar{p})\right\}]$$

　　　　　　　　To facilitate interpretation of the system (14) it is convenient to employ the artifice of a set of four "in effect" singly refractive, transparent objects, using (13) and the ad hoc rules. For each of the four holograms, H_a, H_b, H_c, and H_d, we can invoke such an object, with the optical thickness at an representative location $\bar{\rho}$ on the cross-section given by the values in Table 1.

H_n	$A'_{o_n}(\bar{x},t)$	$\Delta p_n(\rho) = \frac{\lambda}{2\pi}\left[\phi_n(x) - \phi_0(x)\right]$
H_a	$A'_{o_1}(\bar{x},t)$	$\frac{\lambda}{2\pi}\left[\phi_{12_1}(\bar{x}) - \phi_0(\bar{x})\right]$
H_b	$A'_{o_2}(\bar{x},t)$	$\frac{\lambda}{2\pi}\left[\phi_{12_1}(\bar{x}) - \left\{\delta_1 - \delta_{2_1}(\rho)\right\} - \phi_0(x)\right]$
H_c	$A'_{o_3}(\bar{x},t)$	$\frac{\lambda}{2\pi}\left[\phi_{12_2}(\bar{x}) - \phi_0(\bar{x})\right]$
H_d	$A'_{o_4}(\bar{x},t)$	$\frac{\lambda}{2\pi}\left[\phi_{12_2}(\bar{x}) - \left\{\delta_1 - \delta_{2_2}(\bar{\rho})\right\} - \phi_0(x)\right]$

Table 1. Optical thickness of the four "in effect" singly refrac‌tive objects associated with the third prototype hologram.

　　　　　　　　Let us consider the object associated with the hologram (13) is the superposition of this set of four non-obstructing element, such as depicted in Fig. 9.
This figure summarizes the facts that for each exposure, m = 1,2,

each of the two orthogonal polarizations, P_m, gives rise to an independent "in effect" object; the two surfaces α_n and β_n of these four objects n = 1,2,3,4 are depicted there, symmetrical to the position of the unstressed actual object. The arrow num bers relate the center column of Fig. 9 to the six phase pairs of (14). For example, the two arrows labeled (4) indicate that the phase comparison of in effect objects n = 2 and n = 3.

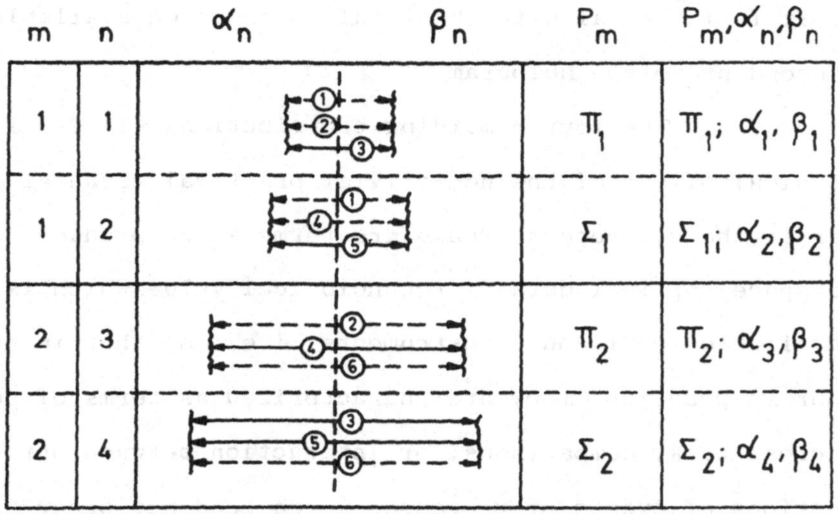

m	n	α_n	β_n	P_m	P_m, α_n, β_n
1	1			π_1	$\pi_1; \alpha_1, \beta_1$
1	2			Σ_1	$\Sigma_1; \alpha_2, \beta_2$
2	3			π_2	$\pi_2; \alpha_3, \beta_3$
2	4			Σ_2	$\Sigma_2; \alpha_4, \beta_4$

Fig.9 - Decomposition of a Representative Thickness Element, say at $\bar{p}$, of the Model, into the Four Associated "in effect", Singly Refractive, Non-Obstructing Specimens, $[P_m, \alpha_n, \beta_n]$.

The fringe system contribution, $F_1(\bar{x}, \bar{x}_0)$ in $F(\bar{x}, \bar{x}_0)$ due to (14.1) relates to the birefringence component of the optical property of the actual model, under conditions of the initial prototype hologram. The presence or absence of an applied load, or residual stress-induced birefringence is indicated by this datum.

The contribution $F_6(\bar{x}, \bar{x}_0)$ in $F(\bar{x}, \bar{x}_0)$ is the isochromatic fringe information characteristic of the terminal load. It is identical with the total information available from the second prototype hologram.

The four remaining contributions (14.2 - 14.5) are typical of one of the non-trivial practical advantages peculiar to the HI process. These are terms which produce, in the image space, optical data of technological value which is often not available in CI and P instruments, i.e., neither in principle nor in practice. They are characterized as terms of coupling, cross-talk, comparisons, or interaction between optical inspections of events or processes which need not occur simultaneosly in time, but merely nearly identically in three-dimensional space. The first of these, $F_2(\bar{x}, \bar{x}_0)$, quantitative compares the singly refractive local thicknesses of the model for the two conditions of loading, the isopachic datum.

It may be recalled that the effective composite object associated with the HI analysis of a vibrating surface is an artifice which experimentaly acquired sequentially (i.e. in series), the extreme positions are recorded serially, respec ively, and thus non-simultaneously. When the "object" was a Michelson interferometer [8], the effective object was experiment- ally composed by either sequential (series) or simultaneous (par allel) operations. In this sense, our third prototype hologram experimentally composes the four effective objects by a hybrid series-parallel process. In each of these cases, the reconstruc- tion process interrogates the set in parallel, in accordance with the equivalence principle discussed in the Powell and Stet- son papers [7] , [8] ,

3.5. Some Recent Contributions and Speculations

A certain number of authors have been taken part at holography work last years, giving contributions not only to photoelasticity but also to other branches of experimental stress analysis.

Nicolas [15] used a special material, polyester resin Stratyl A16, having the property of changing its photo--stress constant sign with temperature. Thatway it has been poss ible by combining a series of four holograms (in fact two double exposure holograms) to make the optical path practically independent of changes of refractive index due to photoelastic effect. Thatway, the result of interference at reconstruction process have been only isopachics, the loci of points of constant sum of transverse principal stresses, as the registration of the change in the model thickness.

Fourney [11] gave attention to another possibility of separating of principal stresses by applying the method of oblique incidence in photoelasticity. Since the hologram record ed three-dimensional image of the model, the normal and oblique incidence fringe pattern may be obtained from a single hologram. Fig. 10 represents a schematic view of the arrangement for obtaining normal and oblique incidence stress pattern. These fringe patterns could be obtained from both real and virtual image,

and the angle of rotation may be selected at any position from
a single hologram.

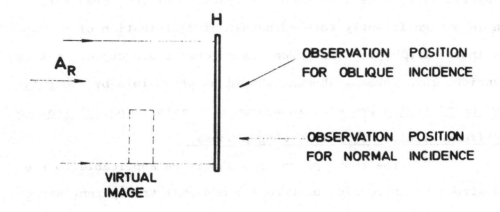

*Fig.10 - Arrangement for Obtaining Normal and Oblique
Incidence Stress Pattern.*

          Up to now we have considered the HI apparat
us in application to problems which have been specialized severe
ly, namely, to the study of two-dimensional, plane object problems. One advantage has been that the analysis is readily manage
able. Even with this degenerate problem study, the results already imply some distinct practical values of the instrument.
It may be in order to wonder about possible more generalized
engineering uses of the process.

          In connection with plane problems, the avail
ability of $F_3(\bar{x}, \bar{x}_0)$, F_4, and F_5, (see Eqs. 14), together an

analytical understanding of their meaning may be of more than academic interest to application enginners. The question of a non-destructive three-dimensional inspection of the local situation in a significantly three-dimensional distribution of refrac tive index $\bar{\bar{n}}(\bar{p})$, i.e., a three-dimensional phase object, entices attention. This accomplishement should be attanaible by the <u>pola rization filtering process</u>, in conjunction with a <u>spatial frequen cy difference (i.e. $\bar{\omega}_{ij}$) filtering process.</u>

The above procedures should be applicable to the nondestructive inspection of three-dimensional transparent struc tures and the properties of their materials, with as well as without, the aid of accessory influences such as the $I(\bar{k})$ of Fig. 4. Conversely, with material $\bar{\bar{n}}$, with a known local optical response to I, the HI process should offer the basis for develop ment of intruments which sense, or moreover, analyze the influence $I(\bar{k})$.

Considering the growing sophistication of laser technology, c.w. pulsed modes of operation, internal and extern-al control of temporal coherence, the hologram interference process should be useful in the study of static, temporally pe-riodic, transient and stochastic processes and events which can be correlated with refraction index distribution $\bar{\bar{n}}(\bar{p})$ and/or external influences $I(\bar{k})$.

For the time being, there has been a series of

papers dealing with different dynamical problems by using hologram interferometry. The pioneers in this field are Stetson and
Powell [6], [7], [8] by their contributions of hologram interferometry in inspecting vibration phenomena, showing the possibility of separate controlling of both frequency and amplitude of
excitation. The hologram interferometry fringes which appear in
the wavefront reconstruction are the mapping of amplitudes,
nodes of vibrations and displacement distribution.

      The application of HI at higher frequencies
problems is also very significant, due to the existence of puls
ed lasers which may be used at wavefront recording process too.
If a pulsed laser is used at the recording process, then the
first step requires a period of the order of a few hundred
milliseconds or several nanoseconds. Even for such intervals,
however, multiple exposure interference effects can make visible
certain refractive index change events. This aspect of the wave
front reconstruction techniques has been pioneered by Brooks
and his associates [18]. By properly synchronizing the trigger-
ing of a laser pulse and the fight of a bullet, evidence of the
shock wave in the neighborhood of the bullet has been made visible.

      Any object which experiences a discrete or
vibratory mechanical strain may be inspected by the HI techniques.
The element to be studied need not be modified by polishing, by

addition of reflective or dichroic coatings; it need only to be
viewable from the hologram plate. The magnitude of displacements
which can be detected is limited primarily by the size of wavelength of the coherent illumination.

The additional possibilities to the extension of
the strain and stress analysis are the applications of holography
to photoelastic coating techniques and to Moiré method.
Recently, a paper by Sciamarella[19] appeared, discussing two prob
lems associated with the practical application of Moiré fringes:
the first - obtaining of sufficient sensitivity for measuring
small strains, and the second - developing of a simple and inex-
pensive technique for engraving lines on the surface of a model,
necessary to the application of Moiré method. The technique pre
sented in this paper combines the idea of diffraction-order filtering with two beam interferometry method.

In the paper by Gottenberg[20] a very exhaustive
description of the application of hologram interferometry in
strain analysis has been given. The author inspected as the bend
ing problem of a prismatical bar using a continuous 15 mw gas
laser as the wave propagation problem in a bar by using pulsed
laser having a power output of 10^7 mw over a time interval of
0.1×10^{-6} sec.

Due to tremendously developing techniques of new
types of both continuous and pulsed lasers the most general prob

lem of the determination of the displacement state on the surface of a three-dimensional body is the possibility to the direct
application of hologram interferometry at both static and dynamic
cases.

REFERENCES

[1] D. Gabor, "A New Microscopic Principle", Nature, 161, 777, 1948.

[2] E. Leith and L. Upatknies: Jnl. Opt. Soc. Am., 52, 1123, 1963.

[3] E.N. Leith, A. Kozma, J. Upatknies, J. Marks, and N. Massey, "Holographic Data Storage in Three-Dimensional Media", Appl. Optics, Vol. 5, 1303, August, 1966.

[4] G.W. Stroke, " An Introduction to Coherent Optics and Holo graphy", Academic Press, New York-London, 1966.

[5] Mme Pauthier-Camier, "L'Holographie", Rev. Franc. de Mec., 26, 89-95, 1968.

[6] R.L. Powell, K.A. Stetson, "Interferometric Vibration Analysis by Wavefront Reconstruction", J.Opt. Soc. Am., 55 (12), 1593, Dec. 1965.

[7] K.A. Stetson and R. L. Powell, "Interferometric Hologram Evaluation and Real-Time Vibration Analysis of Diffuse Objects", J. Opt. Soc. Am. 56 (12), 1694-95, Dec., 1965.

[8] K.A. Stetson and R.L. Powell, "Hologram Interferometry",J. Opt. Soc. Am., 56 (9), 1161-1166, Sept., 1966.

[9] B.P. Hildebrand, K.A. Haines, "Interferometric Measurements Using the Wavefront Reconstruction Technique", Appl. Opt.,5 (1), 172-173, Jan., 1966.

[10] K.A.Haines, B.P. Hildebrand, "Surface Deformation Measure-
     ment Using the Wavefront Reconstruction Technique", Appl.
     Opt. 5 (4), 595-602, Sept.,1966.

[11] M.E. Fourney, "Application of Holography to Photoelastici-
     ty", Exp. Mech., 33-38, Jan., 1969.

[12] J.D. Hovanesian, V. Brcic, and R. L. Powell, "A New Stress-
     -Optic Method: Stress-Holo-Interferometry", Exp. Mech.,362-
     -368, August, 1968.

[13] V. Brcic, "Hologram Interferometry and Its Application to
     Experimental Stress Analysis", Transactions, Inst. Jar.
     Cerni, Belgrade, 43, 26-30, 1967.

[14] R.L. Powell, J.D. Hovanesian, V. Brcic, "Hologram Interfero
     metry with Birefringent Objects" (unpublished text),1968.

[15] J.R. Nicolas, "Contributions à la détermination de la somme
     des contraintes principles ay moyen de l'Holographie", Rev.
     Franc. de Mech., 28, 48-56, 1968.

[16] D. Gabor, "Microscopy by Recontructed Wave-fronts", Proc.
     Roy. Soc., London, A 197, 454-489, 1949.

[17] E.N. Leith and J. Upatknies, "Reconstructed Wavefronts and
     Communication Theory", J. Opt. Soc. Am., 52 (10)

[18] R.E. Brooks, L.O. Hoflinger, R.F. Wuerker, and R.A. Briones,
     J. Appl. Phys., 37, 642, 1966.

[19] C.A. Sciamarella, "Moiré -fringe Multiplication by Means
     of Filtering and a Wave-front Reconstruction Process", Exp.

Mech., 179-185, April, 1969.

[20] W. A. Gottenberg, "Some Applications of Holographic Inter-
      ferometry", Exp. Mech., 405-410, Sept. 1968.